D0457823

PAUL S. SCHUDER
Science Collections

Woodland Public Library

No Bones"

A Key to Bugs and Slugs, Worms and Ticks, Spiders and Centipedes, and Other Creepy Crawlies

by Elizabeth Shepherd
illustrations by Ippy Patterson

MACMILLAN PUBLISHING COMPANY
NEW YORK
COLLIER MACMILLAN PUBLISHERS
LONDON

The author wishes to acknowledge her debt to Miss Alice Gray of the American Museum of Natural History for her careful critical reading of the text and her helpful review of the illustrations. She is also grateful to Russell Frost III for his indispensable contribution.

Macmillan Publishing Company
866 Third Avenue, New York, NY 10022
Collier Macmillan Canada, Inc.
First Edition
Printed in the United States of America

10   9   8   7   6   5   4   3   2   1

The text of this book is set in 13 point Bauer Bodoni.
The illustrations are rendered in pen-and-ink.
Library of Congress Cataloging-in-Publication Data
Shepherd, Elizabeth. No bones. Summary: An introduction to the world of invertebrates and a key to their identification. I. Invertebrates — Juvenile literature. 2. Invertebrates — Identification— Juvenile literature. [1. Invertebrates. 2. Invertebrates — Identification I. Patterson, Ippy, ill. II. Title.
QL362.S537   1987   592   87-1549
ISBN 0-02-782880-8

To my mother

CONTENTS

When you see two pictures of the same animal side by side, the smaller drawing is about the size the animal really is. The larger drawing shows how the animal might look under a hand lens, or magnifying glass.

SMALL ANIMAL TRACKING

What animal has six wiry legs and runs across sidewalks and sandwiches?

It is small. It has long feelers on its head. It has a hard, shiny outside.

It has no leg bones or back bones. It has no bones at all.

Do you know what it is?

It is an ant.

You can find ants almost anywhere. Ants run about in fields and playgrounds, in houses and junkyards, up trees, across rocks, into sugar bowls. In winter you can find them under stones or behind the bark of dead trees or in the ground.

Many other wild animals live in such places, too. Some can fit into small, tight spots. Others move about in big, open spaces.

This one may be feeding inside a soft green plant.

This one may be resting on a door-step.

Like ants, these animals have no bones. They have no leg bones, no back bones, no bones anywhere in their bodies. But these animals are not ants. What are they?

The key in this book will help you find out. Before you start tracking animals, learn how the key works.

Try the key first with an animal you know. This may be the ant or any other animal that lives on land and has no bones.

Start on page 6, where the key

begins. You will find two statements there, both with the number "1" beside them. One statement tells something that is true about your animal. The other does not.

Look at your animal. Read each statement. Pick the one that fits your animal best. Then turn to the number given for the next clue. When you come to the last clue for your animal, you will be given the page number of its story. You may see the animal's picture there, too.

If you want to find out more, one of the books listed on page 86 may tell you what you want to know. On land, there are thousands of kinds of animals without bones. There are millions more living in water. No one book can hold them all.

Because there are so many kinds of boneless animals, you do not need to go far to track them. You need only to look and to listen. Is that flower bending all by itself? Can that tree bark be moving? What is making that buzzing sound?

When you see an animal, you must move very slowly. You do not want to scare it away. Do *not* rush to catch it, either. You can easily hurt a small, soft animal. And some can hurt you.

An ant has no teeth. Yet it can bite—and some kinds just may bite you. A wasp can sting. This sign ⊘ means, Look out! Do not touch!

To be a good animal tracker, you do not need to touch the animals. Watch one as it rests or feeds. It may even rest on you. Think about what it is doing and how it might get away from you. Then, slowly, take out the book and track the animal in the key.

If you wish, you can catch the animal. Just be sure that it will not fight back. You may use your hands or a net. Sometimes you can trap an animal with a baby-food jar. The animal may walk right in. Or you can set the jar over the animal. Once the animal is inside, quickly tip the jar up and screw on the lid.

Be sure to keep the jar in the shade, so that the animal does not get too hot. But do not worry about its getting the air it needs. A small, quiet animal can get enough air to live for many hours even in a small baby-food jar.

For looking at your animals, a hand lens will help. Move the lens slowly *away* from the animal, so that you do not crush it by mistake. Of course, the animal may not hold still while you look. If that is the case, put it in a cool place for an hour. The cold will slow the animal down.

Also, put only one animal in each jar. If two are together, one may eat the other one.

Once you know the name and story of your animal, you should open the jar and let it go. If you can, put the animal back where you found it.

Now you are ready to go tracking again.

A KEY TO ANIMALS
WITHOUT BONES

Key 1   **1**   The animal has legs. Go to 2.

            1   The animal has no legs. Go to 10.

          This animal has legs.

          This animal has no legs.

Does your animal have legs? Choose the one statement that tells about your animal. Then turn to the next clue.

2 Animal has ten or more legs. Go to 3.

2 Animal has less than ten legs. Go to 6.

Key 2

3 Animal has exactly 14 legs. Turn to page 21 for its story.

Key 3

3 Animal has more than or less than 14 legs. Go to 4.

Key 4   4️⃣ Animal has legs on each body ring. Go to 5.

   4 Animal does not have legs on each body ring. *Turn to page 82 for its story.*

What is a body ring? To find out, turn to page 15.

5 Animal has a flat body with legs that stick out at the sides. *Turn to page 23 for its story.*

Key 5

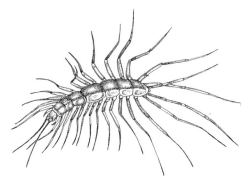

5 Animal has a tubelike body with many short legs close to its sides. *Turn to page 25 for its story.*

Key 6

6 Animal has eight legs. Go to 7.

6 Animal has six legs. *Turn to page 52 for the story.*

Key 7

7 Animal has a body with only one part. Go to 8.

7 Animal has a body with more than one part. Go to 9.

8 Animal has very long, stiltlike legs. **Key 8**
Turn to page 29 for the story.

8 Animal is tiny and its legs are tiny.
Turn to page 31 for the story.

Key 9

9 Animal has big claws and a long tail. *Turn to page 27 for the story.*

9 Animal does not have big claws and a long tail. *Turn to page 33 for the story.*

10 Animal has a shell. *Turn to page 16 for the story.*   **Key 10**

10 Animal has no shell. Go to 11.

11 Animal has feelers. *Turn to page 16 for the story.*   **Key 11**

11 Animal has no feelers. Go to 12.

Key 12   🔟 Animal has more than 15 body rings. *Turn to page 19 for the story.*

🔟 Animal has less than 15 body rings. *Turn to page 80 for the story.*

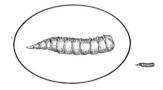

What is a body ring? To find out, read on.

Perhaps you have played ring toss. At the end of the game, you stack the rings on the stick.

The body parts of some animals look like rings on the outside. In any one animal the rings are about the same size and color. They are much the same inside, too.

You can see the body rings on this animal.

There are no rings on this one.

Please go back to your place in the key.

15

SNAILS AND SLUGS

Snails have shells and most **slugs** do not. In other ways they are much alike. Their bodies are mostly head and foot. (This is also true of the kinds that live only in water.)

You can tell the head end from the foot by the head's four tentacles (TEN·TA·KULLS), or feelers. The snail feels its way with its tentacles. So does the slug. Through damp leaves, up and down living plants they go.

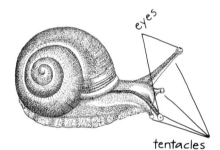

eyes

tentacles

There are two long tentacles on top of the head and two shorter ones close to the ground. The snail has eyes at the tips of the longer ones. So does the slug.

A snail can turn its tentacles up and down, over and under. It can turn them outside in. It can pull them back inside its head. So can the slug.

If a tentacle gets bitten off, the snail can grow a new one. It can grow a new eye. So can the slug.

Of course, the snail can pull itself inside its shell— or most of itself. A slug can pull its head into its "neck."

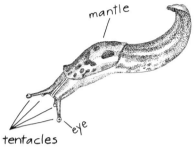

mantle

tentacles

eye

To test this, try tickling a slug. Its head goes under its mantle (MAN·TILL). The mantle is a soft hood of flesh that covers most of the slug's back. It also acts like a lung. If you look carefully, you can see a small hole on the right side of the mantle. It opens and closes as the animal breathes.

A mantle covers the snail's body, too. It makes the shell grow as the snail's body grows.

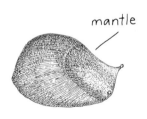

mantle

When the snail or slug moves along, a sticky, milky stuff pours out of the foot. This slime shines on the earth or on a leaf. The animal moves away, but the sticky slime stays.

The slime lets snails and slugs glide over rough ground. It keeps them cool and damp on a hot day. It may keep you from picking up a slug.

You can use your hand lens to scoop up

a slug. Of course you can pick a snail up by its shell. Once the slime starts coming out, it helps the animal stick to the lens. Then you can tip the lens and watch the bottom of the foot moving.

As the snail or slug creeps across the lens, you can see a dark stripe along the foot. This stripe is muscle (MUSS·EL). It shows through the thin skin.

Snails and slugs move about very little during the day. Because they have such dull colors, they are hard to see against dead leaves and rotting wood.

After a rain you may find these animals in the open. Land snails and slugs need water, but not too much all at once.

EARTHWORMS

Earthworms are long and cool and slimy. They go over the ground by night, under the ground by day. They eat their way along. In go bits of seeds, stems, leaves, and soil. Out come coils of leftovers.

head end

To crawl about, an earthworm stretches out long and thin. Then it pulls up short. Tiny bristles (BRISS·ILS) on each body ring keep it from slipping back. The bristles are hard to see, but you can feel them. Just slide the worm over your hand.

Be very gentle or the worm may break. If it does break, lay the *front* end down carefully. This end can grow new rings and become a whole worm again. The back part cannot do so.

This may seem odd, because the front part looks dead and the back part is wiggling. The same thing happens when a bird catches a worm. The bird eats the wiggling part. The front end can crawl safely away.

In winter earthworms stay deep underground, often curled up with other earthworms. They do the same if the summer is hot and dry. Cool spring and fall days are best for finding earthworms. They may even stick their head ends or their hind ends out of their tunnels into the air. (They jerk them in quickly when they feel you coming.)

Sometimes you may find many earthworms together in a pile of wet leaves. When you do, you may also find their egg cases. The gray-brown cases look like pale seeds against the dark leaves. About the size of this O, each case holds many eggs.

Put a few cases aside in a jar of damp soil under some leaves. In a few days the eggs will hatch, and tiny earthworms will crawl out of the cases. When you let them go, it may take them awhile to crawl out of sight.

ISOPODS

People call these crusty little animals all sorts of names. **"Isopod,"** "sow bug," "pill bug," "roly poly," "water bug," "wood louse," and "slater" are some of the names.

Some, like the roly poly, can bend into a U shape. Other isopods (I·SO·PODS) cannot.

Either way, an isopod has a small head with long feelers and a short, wide body with 14 legs. If your animal has short feelers and more than 14 legs, turn to page 25.

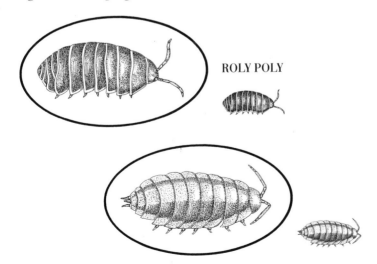

ROLY POLY

An isopod has strong jaws. It can tear into a soft plant stem or root. It can grind down tough leaves. It feeds on live plants as well as dead ones. It eats dead animals, too.

The mother isopod carries her eggs in a sort of pouch under the front end of her body. When the young hatch, they start right in feeding.

Young isopods look like their parents except for their tiny size. Gray or brown in color, they match the color of the soil or tree bark or dead leaves where they live.

CENTIPEDES

How many legs do **centipedes** (SEN·
TI·PEEDS) have? They seem to be all
legs and feelers. The name means "one
hundred feet." Most centipedes have 30
legs, but some have as
many as 250 or more.

Each body ring has two
legs. The legs stick out at
the sides. As the centipede
walks, it swings its body
from side to side. That
way the legs do not step
on one another.

Centipedes are hard to
catch. They can run fast.
Their flat bodies can slide
into cracks and other tight
places.

feelers

eyes

Often you find a centipede under
stones or rotten pieces of wood. When
you lift up its home, the centipede runs
this way and that. It does not see you,
but it can see the light. It runs until it
reaches another dark place. Centipedes
seem to feel safe when their bodies
have something to touch on every side.

23

Centipedes rest all day and come out at night to go hunting. They hunt mostly by touch. As a rule, they eat insects and other small animals.

Fast and fierce, house centipedes can catch big water bugs. Other centipedes can kill frogs, toads, and even mice. They can do this because they use poison.

A centipede's first pair of legs are part claw and part poison fang. The centipede grabs its prey and shoots in the poison. Then it can eat.

Most centipedes are too small to hurt humans. A few large ones can and will bite if you scare them. Their poison hurts, but it is not strong enough to kill you.

MILLIPEDES

Millipedes (MILL·LI·PEEDS) have *four* legs on each body ring. The legs are short and slow, but strong. A millipede uses its legs in pairs almost as though swimming. It can climb up a stem. It can tunnel into soft earth or push into a crack. Always it keeps out of the sun.

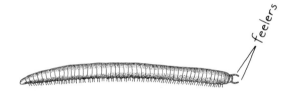

The wireworm-millipede is long and thin. The pill-millipede is short and plump. Long or short, all millipedes are slow. Their food does not run. Millipedes eat plants—very young ones or dead ones or sick ones. They eat dead animals, too. As they move along feeding, their hard outsides keep them from harm.

To get away from danger, most millipedes just curl up. They lie on their sides and coil their bodies around their heads.

The pill-millipede presses its front and back ends together. It looks something like a roly poly (see page 21), until you count the legs.

Sometimes mother millipedes also curl up around their nests. They guard the eggs until they hatch.

No one knows if the first millipedes did this millions of years ago. No one even knows exactly what they looked like. But they did make the same sort of tunnels as they do today. Most likely millipedes then lived very much as they do now. As far as anyone knows they were the first animals ever to live on land.

SCORPIONS

⊘

A **scorpion** (SCORE·PEE·ON) holds its long tail over its back or off to one side. It is always ready to sting.

The sharp tip of the tail is full of poison. The poison of this brown scorpion can kill people. The poison of other scorpions just hurts—but that is bad enough!

Scorpions live in places where winters are warm. By day they hide under stones, down holes, or in rubbish piles. By night they hunt.

A scorpion runs on eight legs with its two huge claws way out in front. The claws act as feelers—and as grabbers. If the prey is strong or wiggly, the scorpion stings it. It mashes up its food with the small claws in front of its mouth. Then it sucks out the soft parts.

Scorpions feed on spiders and centipedes. They feed on beetles and other insects. Some large kinds even eat mice. All scorpi-

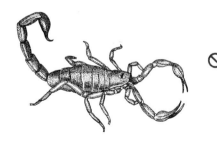

⊘

ons can go many days without food and water.

With food or without it, scorpions live alone most of the time. But a mother scorpion takes care of her young after they are born. She carries them around on her back until they are old enough to hunt their own food.

DADDY LONGLEGS

Daddy longlegs skitter across dead leaves and over living plants. Because so many crowd together at harvest-time, they are also called harvestmen.

In the South they may also spend the winter in these large groups. In the North most daddy longlegs die as days and nights grow cold. Their eggs hatch in the spring, and soon you can find daddy longlegs running about again.

Their long, skinny legs bend out and over and under their gray-brown bodies. The two longest legs feel the way. They reach up, back, and all around. When in danger, a daddy longlegs may cast off a leg, then run away. If the enemy stops to look at the leg, perhaps the daddy longlegs can escape.

Somehow a daddy longlegs can get along without all eight legs, for it cannot regrow a lost one. It can live with as few as four legs. This is true as long as it has *one* of the longest two left.

A daddy longlegs feeds on dead insects—and on live ones. It also eats tiny spiders and snails. Its short, hairy feelers first touch, then hold the food. Its mouth claws are sharp but very small. With a hand lens you can just see them.

After it eats, the daddy longlegs cleans up. It pulls each leg through its mouth claws. It scrapes dirt off each feeler the same way.

Sometimes you see a dead daddy longlegs hanging in a spiderweb. The spider does not eat it. Daddy longlegs have stink glands on their front legs. After one taste, the spider runs off and wipes its mouth on a leaf.

MITES

Mites often find you before you find them. In fact, most mites are so small you can hardly see them. Many spider mites stand out because of their red color. They feed on green plants.

To find other mites, look carefully at the undersides of leaves. Mites often spin a tangle of silken threads from leaf to leaf. This lets them run quickly between the leaves—and it helps you find them.

TICK

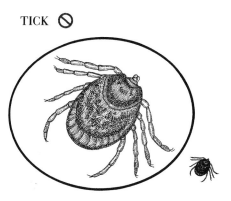

You can also find them in another way. Hold a sheet of white paper under a leaf and shake or tap the leaf. The mites—if there are any—will fall onto the paper. Aphids (see page 70) and other tiny animals may fall also. Some are food for mites of many kinds. In turn mites are food for still other mites, for insects, and for spiders.

Ticks are the largest mites. They look like brown spots on a leaf—until the right animal brushes past. Then the tick grabs hold of the animal with its front legs and jumps on. It may walk around for a while hunting for bare skin.

At last it jabs in its beak. The other animal does not feel it. Then milky glue pours out and holds the beak in place. The tick slowly sucks in the animal's blood or other body juices.

The tick may feed for two or three days. Then it drops to the ground almost too fat to walk. It hides then and sheds its skin. If a female, the tick may now lay her eggs, becoming small and thin again.

By this time the bite on the animal starts to itch. The bites of chiggers and some other mites itch for many days. They may also carry germs that make people and other animals sick.

SPIDERS 🚫

Spiders are shy. They run from you if you get too close. They hide in a crack or under a leaf or a stone. The house spider may put its legs in front of its face and seem to hide.

When a spider cannot run away, it may bite. Of course, most spiders are too small to prick your skin. Some kinds are no bigger than a period. Others *are* big enough. A few kinds have a poison that can make you very sick.

So, do not try to trap or hold *any* spider. Instead, watch them as they wait in their webs or sit on a leaf. You can use a hand lens to watch them.

Most spiders have eight eyes. But few spiders can see very well or very far. They cannot see you if you move slowly.

Spiders do not have ears or noses. They do have many hairs all over their bodies and legs. Each hair is like a mini-feeler.

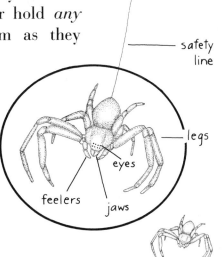

safety line

legs

eyes

feelers

jaws

33

Spiders sense the world mainly through their hairs.

The feelers themselves are very, very hairy. They look like very short, thin legs at the front of the body. The feelers of the male have knobby tips.

The spider's mouth lies between the feelers. It is behind the hairy jaws. The base of each feeler is heavy and thick with sharp edges. The two feelers work together, crushing and holding the prey.

A spider has poison fangs at the tips of its jaws. The fangs of a tarantula (TAR·AN·CHEW·LA) jab straight down and drag in the prey. The fangs of the house spider pinch together like one big claw.

Spiders often catch animals larger than themselves. They can do this by using their poison and by using their silk. A house spider, for example, throws silk all around a fly. It ties the fly up and then gives it a shot of poison. The poison makes the prey quiet. The spider may wrap it up some more—just in case. It can then have its meal without any more trouble.

Spiders have tiny mouths. Most cannot chew their food. Their saliva (SAL·I·VA) makes it soft, so they can suck out the soupy juice. Many mother spi-

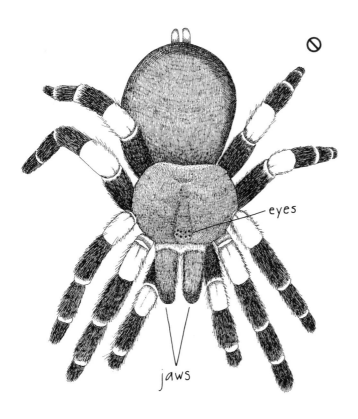

eyes

jaws

ders share their food with their young. If a spider is not hungry, it wraps the food with silk and saves it for later.

Toward the end of summer, when spiders are fully grown, their webs are often large. You can find them in all sorts of places. Some spiders put them high in the trees. Others spin them in animal holes or across windows. When you see *where* the web is, you can often guess *what* it is likely to catch.

Spider silk is so fine that it is hard to see. It is easy to walk into a web. If you have ever walked into one, you know how strong it is and how stretchy.

If you find a big web, look carefully at the spider. As a rule, it will be a female. To make sure, look at the feelers. Remember: The tips of the female's feelers look like feet; the male's look like tiny fists.

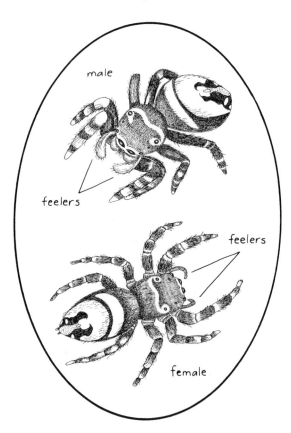

male

feelers

feelers

female

Hold a hand lens over the belly of your spider. If it is big, you may be able to see the tiny tubes called spin-nerets (SPIN·NER·ETTS). The spider may use its back legs to pull silk from the spinnerets. The spider pulls out the

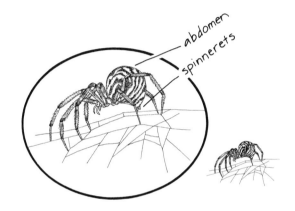

abdomen
spinnerets

milky silk, first with one hind leg, then with the other. The silk stretches into threads.

Often a spider just presses its spin-nerets against a branch. The silk sticks there like a dab of glue. Then as the spider moves, the silk flows out in a strong thread.

Each kind of spider makes its own sort of web, or it may make no web at all. Tarantulas grab crickets when they feel them pass. They do not use a web. But tarantulas can and do spin silk.

All spiders wrap their eggs in silk. They may mix in body hairs to make a thick, warm case. Some spiders make flat covers over their eggs. Others make round balls. Some hide their egg cases. Others carry the case around with them.

egg case

As soon as a spider hatches, it can spin. It climbs up and up, as high as it can go. A thin wisp of silk trails out into the breeze. The spider lifts off. It blows with the breeze like a tiny balloon.

When the wind stops or a branch snags the silk, the spider runs down. It

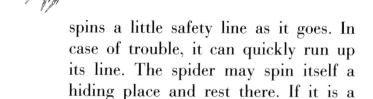

spins a little safety line as it goes. In case of trouble, it can quickly run up its line. The spider may spin itself a hiding place and rest there. If it is a web spider, it may start a web.

As a spider grows, its tough skin does not grow. The spider must *molt*, or shed its old skin. It grows and molts, grows and molts. Its body always has two parts.

When a spider is molting, it cannot bite or run. It may spin a nest to hide in. Even so, a bird may find it, or a toad—or another spider.

Certain wasps are any spider's worst enemy. A wasp stings the spider and carries it home to feed to her young. If the spider is big, the wasp may dig its nest around it.

Even if they are not eaten, most spiders live only one year. Some wolf spiders may live two or three years. A few pet tarantulas have lived about twenty.

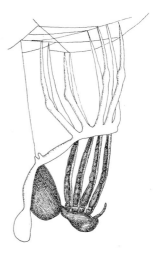

MINI-KEY
TO SPIDERS

Mini-Key a   [a] The spider has a web. Go to b.

[a] The spider has no web. Go to e.

Mini-Key b   [b] The spider makes a web that looks like a stop sign. Read below.

[b] The spider does not make a web that looks like a stop sign. Go to c.

☐ Many spiders spin webs that look like the ghost of a stop sign in the air. If you find a stop-sign web, it is likely that the spider will be female. The males are smaller and spin much smaller webs.

To see the spider spinning, go back to the same place just before dark or, better yet, very early the next day. The spider will make a new web almost every day.

Male and female both follow the same plan. First, they spin a strong

40

frame. Like a bicycle wheel, the frame has many spokes. Then the spider spins a path of silk, around and around, toward the center. This path is sticky. It catches insects. The frame silk is dry. It does not catch the spider. The spider knows which kind of silk is which, but a fly does not.

Some big spiders "sign" their webs with a zigzag of thick, white silk. The black-and-yellow garden spider is one of them. She waits in the center of her big web, her legs making an X against the zigzag. Maybe the zigzag keeps birds from flying into the web.

Many smaller, thickset spiders sign their webs with a long, thin line. They do not sit in their webs, but a long line leads your eyes on. There the spider hides in a rolled-up leaf. Still other spiders leave an open space at the center of their webs. There they sit, quiet as seeds or bits of leaf.

Mini-Key c $\boxed{\text{c}}$ The spider spins a web like a smooth sheet. Read below.

c The spider does not spin a web like a smooth sheet. Go to d.

☐ Many spiders spin thick, smooth sheets of silk. They spin them across grass or tall weeds, over walls, or high up in the trees. The webs shine with dew on summer mornings.

In a way, these webs are like circus nets. A grasshopper may jump in. But when it tries to jump out, it sinks into the net.

Look carefully at each web. The web may be flat as a mat or round as a bowl. The spider may be waiting under it. It reaches up through the web to bite.

Some spiders crisscross silk lines above the net. When an insect bumps into the lines, it falls into the net.

42

Grass spiders make a smooth funnel of silk off to one side. The spider hides there. When a beetle falls on the web, the spider runs out and pulls it inside. You can tell a grass spider by its long spinnerets.

---

d The spider spins a maze. Turn to page 45.

Mini-Key d

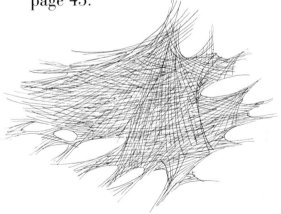

d The spider spins a tube. Turn to the next page.

43

The female purse-web spider spins a tube up the side of a tree. She starts with a small tube on the ground and works her way up. Then down she goes, digging a hole. She pushes sand and dirt out through the sides of her web. This makes the web hard to see. The male's web is much smaller.

When a millipede or isopod walks over the purse, the spider runs up the inside of her tube. She stabs the prey through the silk and then pulls it in. After her meal, she fixes the web. She throws the leftovers out the top.

The tube spider spins a tube in a hole or a crack. It spins silk lines like a collar around the mouth of the tube. Then it sits inside with six of its spiny legs in front. When an animal trips on the lines, the spider is ready to leap.

☐ Many spiders spin mazes of fine silk. The cellar spider spins its mazes in dark, out-of-the-way places. The mother and father spiders share the web. The mother holds the egg case in her jaws. With their tiny bodies and very long, very thin legs, cellar spiders look like daddy longlegs (see page 29).

Watch out if the maze looks very thick and heavy. This is likely the web of the brown, or violin, spider. Its bite

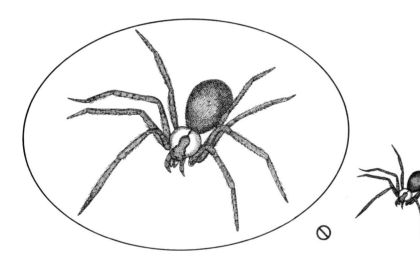

kills the skin around it, and so it takes many months to heal.

A loose, thin maze may mean trouble also—if it belongs to a widow spider. The bite of this spider can sometimes kill a person. The male is much smaller and does not bite.

The widow's cobweb looks much like the web of a house spider. Her body

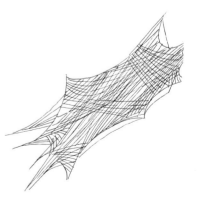

45

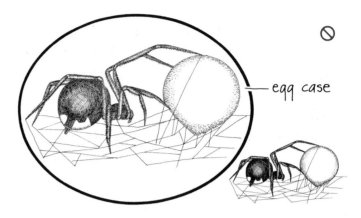

egg case

looks much like the house spider's, but is much larger. Most widows are dark with red markings. A house spider is tan or gray, with black markings. Both kinds hang their egg cases in their webs.

Mini-Key e

e The spider chases its prey. Go to f.

e The spider lies in wait for its prey. Read the next page.

■ The little crab spider can run sideways like a crab. It's short body and spiny front legs look crab-like, too.

Mostly it sits in flowers or on leaves. When a bee or wasp visits the flower, the spider grabs it and bites fast. The crab spider's poison stuns the prey before it can fight.

In the spring many young crab spiders are white. They sit on white flowers. As they grow, they change to yellow. Then they hide on yellow flowers. They can change from yellow to white and back to yellow again. They need about a week to change.

The brown trap-door spider also lies in wait for its prey, but it hides underground. It digs a hole and adds a door of earth and silk. The door fits the hole exactly. The spider lifts the door just enough to grab at insects passing by. Then it slams the door and eats.

Mini-Key f

[f] The spider runs after its prey. Go to g.

■f The spider jumps onto its prey. Read on.

■ A jumping spider sees much better and farther than most spiders. It can see insects as far away as the top of this page is from the bottom.

When it sees an insect, the spider creeps up. At the last second it jumps. Its four front legs are long and strong. They hold the insect tightly until the spider's poison does its work.

During the day you may see jumping spiders walking along a twig or blade of grass. You may find one peeking out of its bag of white silk. Jumping spiders may spend the night inside. They

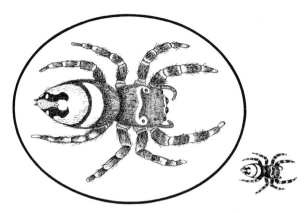

keep their eggs and young inside, too.

A few spiders spit and then jump. The spitting spider sneaks close to its prey. Then it spits out sticky gum. As the insect tries to get free, it gets more and more stuck.

---

g The spider runs about mainly at night. Turn to the next page.

g The spider runs about mainly by day. Read on.

Mini-Key g

■ The wolf spider seems to look right at you with its two largest eyes. In fact, it can see four ways at once. The four small eyes in front look ahead and to the sides. The two large eyes farther back look straight up.

A wolf spider sees very well—for a spider. It cannot see as well as most insects can. But it can run fast and grab hard. Because its legs are so strong, the wolf spider hardly needs poison even to keep a big insect down.

A mother wolf spider carries her egg case on her spinnerets. This leaves her jaws and legs free for hunting. When the young are ready to hatch, she bites open the case and lets them out.

Then the young climb onto her back. She carries them until they molt for the first time.

☐ Many kinds of spiders hunt mainly at night. One of these is the tarantula, which lives in hot, dry places. It has hairy jaws and hairy feelers. It finds its prey by touch.

But even the mean-looking tarantula runs into its hole at any sign of danger. When it cannot hide, it makes itself look meaner. It rises up tall on its hairy legs. It kicks. It throws off hairs. The hairs can hurt the mouth or eyes of a hungry skunk or lizard.

50

And, of course, a tarantula can bite. It can bite hard, for its fangs are big and strong. Its poison is not strong, but it can make some people sick.

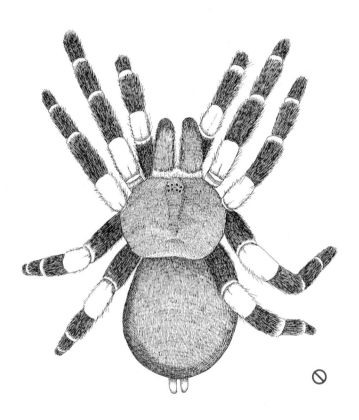

A mother tarantula bites if you touch her eggs. She wraps them in a loose bag of silk. She may take it to a sunny spot to warm the eggs. The rest of the time, she keeps the egg case in her hole. She watches over it until the eggs hatch.

INSECTS

The ant is one kind of **insect.** A beetle
is another. A cockroach is another. All
have six legs.

Some people call any animal with six
legs a "bug." It is true that all bugs are
insects. But not all insects are bugs.
Really, bugs belong to just one of the
many groups of insects. Ladybugs
belong to another group, bees to
another, and so on.

The body of a bug or any other
insect has three main parts. The parts
are the head, the thorax, and the abdomen (AB·DOH·MEN).

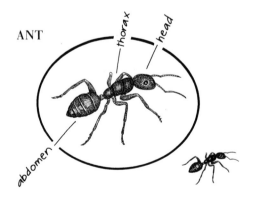

ANT

The thorax has three parts, too. Each part has one pair of legs. Each leg has five joints. The joints let the legs bend forward and backward and to the sides. The feet have claws, which help the insect move over rough places. The feet also have pads and hairs, which keep the insect from slipping on smooth places.

Most grown-up insects have wings. Fleas and some others do not. Male ants have wings for only a short time. Young insects have no wings at all.

FLEA

Can you tell what this insect will be when it grows up? As it grows, its wings grow. Can you tell what it is? (To see if you are right, turn to page 64.)

It is likely that the first insects on earth looked a bit like worms. The young of many insects still look like worms. A young firefly, or glowworm, looks like a worm with six tiny legs. A young weevil looks even more like a worm, since it has no legs at all. (If you want to see what it looks like, turn to page 80.)

The wireworm looks like a millipede with just six legs. It looks nothing like its parent, shown on page 79. A young bean beetle looks like a bit of yellow and white fluff. A young ladybug beetle looks like a mini-alligator.

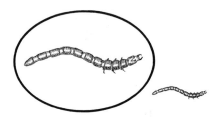

In a few weeks or months, these young insects become *pupas* (P·YOU·PAS). (To find out about pupas, turn to pages 81 and 84.) In time, pupas change into grown-up insects.

Once an insect is grown up, its wings will give you clues to its name or group. Most often, one pair of wings is bigger than the other pair. The veins

(VANES) make the thin wings strong.

An insect's eyes and mouth also give you clues to what it is. Most insects have two large eyes at the sides of their heads and three tiny eyes on top of their heads. You can see both sets of eyes with a hand lens. No one knows for sure the use of the tiny eyes.

With the hand lens you can see the mouth parts. Most insects get their food either by chewing or by sipping. Each kind has its own sort of mouth— and its own kind of best food.

For example, mosquitoes (MOS·KEE·TOES) have a sort of beak. Females can prick skin and suck out the blood, much as you sip soda through a straw. Leafhoppers suck plant juices, or sap. Butterflies sip the juices of flowers.

Crickets and beetles grind their food between strong jaws. They chew from side to side, not up and down as you do.

CRICKET

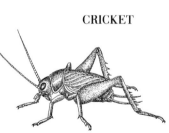

Some insects have ears. In most cases, the eardrums are inside the abdomen, where you cannot see them. Those tiny openings along the sides are for breathing.

An insect has no nose. It does have

antennas, or feelers. Like the legs, they can bend every which way. The antennas smell and even taste food. They sense danger, too.

An insect can sense the world with other body parts. It uses its lower lip and its feet. Some insects have cerci (SER·SEE). The cerci are like feelers at the hind end of the body. The insect can use them backing into cracks.

A cockroach has long cerci. An ant has short ones. An earwig has cerci that can pinch—but not very hard.

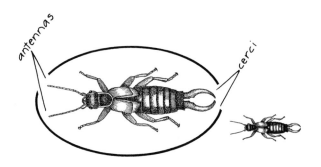

Don't forget! Other insects, too, can pinch or bite or sting. Before you catch an insect, watch it carefully. Look at its wings and its mouth and its legs. You may be able to find out what it is without ever catching it. The mini-key that starts on the next page will help.

MINI-KEY TO
INSECTS WITH WINGS

A The insect has four wings. Go to B. **Mini-Key A**

A The insect has two wings. Read on.

■ Bzzzz! Bzzzz! Tiny wings are beating fast. Mosquitoes are hunting.

A mosquito lands on an animal—on you! You may not feel her. Her mouth parts are finer than needles, and you may not feel them prick your skin. Her saliva (SAL·I·VA) keeps your blood coming long enough for her to suck out a meal. Later, the saliva makes the

"bite" itch. Sometimes it carries germs. Male mosquitoes feed only on flowers.

Many female flies also feed on blood, just like the mosquito, while their mates feed on plants. Others, such as houseflies, do not bite. But they can hurt us, too. They leave germs behind when they walk on our food.

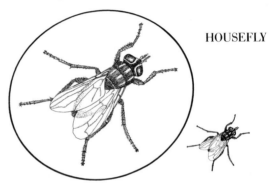

HOUSEFLY

Not all flies are pests to us. Some are pests to other insects. Robber flies catch wasps and beetles and moths. Young bee-flies and flower-flies also eat insects. The grown-ups feed on flowers.

FLOWER-FLY

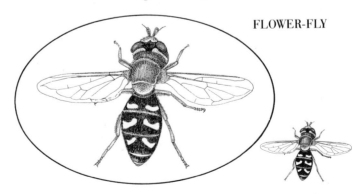

B The insect has "dusty" wings. Read below.

B The insect does not have "dusty" wings. Go to C.

☐ If you catch a butterfly or a moth, its scales dust your fingers. Sometimes you can see scales left behind on a window screen or a spider's web. The scales stick to the web, but the wings may not. Then the insect can get away.

You can easily tell a butterfly from a moth. Its antennas are always straight, with knobs at the tips. A moth's antennas are like feathers or like threads with no knobs.

Both insects sip the juice of flowers. At times their long mouth tubes are rolled up. Many moths never feed at all.

A butterfly flies and feeds in the daytime. As it flies, it shows the pretty side of its wings, then the dull underside, then the pretty side. The changes may fool birds wanting a meal.

A butterfly feeds with its wings up. The colors are hidden. But some butterflies show their stripes and spots. The swallowtail is one. It may feed on flowers whose sap will make it taste bad. Birds and toads soon learn not to eat swallowtails.

Moths feed at night and rest by day on trees and grass. They may hold their wings out or on their backs. Their dull brown and gray colors make them hard to see. Some moths have big spots on their wings. If a bird starts to peck such a moth, the moth opens its wings. When the big spots "stare" up, they may scare the bird away.

At night, moths warm up by beating their wings. Then they take off. The wing scales let them fly softly. Their big antennas help them find flowers—and other moths.

Moths have ears on their abdomens. Maybe they can hear bats coming and so let themselves drop safely to the ground. This moth makes clicks with its hind legs. It seems that bats keep away from the clicks—and the moth.

Mini-Key C

C The front wings are as slim and straight as knife blades. Look at the legs.

> C₁ The front legs are very big and strong. Read below.
>
> C₂ The hind legs are very long and strong. See the next page.

C The front wings are not as slim and straight as knife blades. Go to D.

C₁ The praying mantid has big, strong front legs with sharp spines along their edges. The mantid holds onto a twig with its middle and hind legs.

Its front wings are thin and flat, with many fine veins. The bigger hind wings are folded under them. A male's abdomen is slimmer than a female's.

Most mantids are tan or green. Their color blends with the plants. So do their big, big eyes.

All at once the mantid turns its head. Few other insects can do that. The mantid moves fast and grabs an insect. It holds its prey like an ear of corn. After eating, it cleans its face and legs.

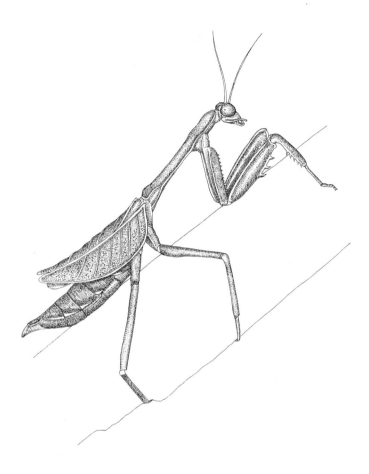

Its cousin, the cockroach, does the same. (See page 72.)

(See page 72.)

C₂ Grasshoppers and locusts are big, noisy insects with long, strong hind legs. The same is true of crickets and katydids. Their slim front wings slope down, green as new leaves or gray and brown as dead ones. Some kinds have long wings. Others have short ones. All have thin heads with flat faces, strong jaws, and long antennas. Crickets may have long cerci, too.

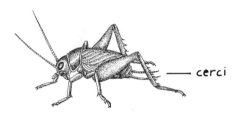
cerci

Most grasshoppers "sing" by rubbing their hind legs against their front wings. Some crickets rub their wings together. The warmer the night, the faster they sing.

When grasshoppers take off, their back wings may snap open like fans. Mostly they leap from plant to plant, chewing big holes in the leaves. When they run out of food in one place, they fly off to look for more.

| D | The front wings are thin and clear as plastic. Go to E. | **Mini-Key D** |

| D | The front wings are not thin and clear as plastic. Go to G. |

| E | The front wings are longer than the hind wings. Go to F. | **Mini-Key E** |

| E | The front wings and the hind wings are about the same size. Read on. |

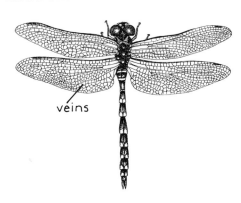

veins

■ Some insects, like the big, bold dragonfly, have four big wings. The dragonfly darts forward and backward all day long. Its head seems all eyes. It hardly needs antennas. It can see ahead and up and down. It can even look backward by bending its head.

The dragonfly cannot fold its wings.

It always keeps them out. It cannot walk. Its legs can hold onto a branch or grab flies and mosquitoes in midair.

The lacewing is another insect hunter with four big wings. By day you may find it on a leaf. It looks just like a leaf when its wings are folded. Only its round, golden eyes give it away.

The lacewing flies at night from one low plant to another. It walks along, eating tiny insects.

Mini-Key F

F When resting, the insect holds its wings up over its body. Skip to page 70.

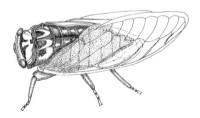

F When resting, the insect holds its wings out or to the sides. Read on.

White and gold, yellow and black are the colors of wasps and bees. The colors warn off snakes and birds and people.

As a rule, bees are heavy and hairy while wasps and hornets are slim and shiny. Like their ant cousins, each has a waist between its thorax and its abdomen.

YELLOW
JACKET

Few people think much about what a wasp or bee looks like. They think about how it stings. The sting hurts! Worse still, it can make some people very sick, or even kill them.

A few ants, such as fire ants, can sting also. Most other ants only bite.

As you watch a bee on a flower, you do not see the stinger. It is inside the abdomen until the bee needs it. The same is true of most wasps and hornets. Only the females can sting.

Some small wasps seem to have very long "stingers" for their size. The stinger is always out. Yet most of these wasps cannot sting. They use the long stinger only to lay eggs. They lay eggs on other insects and on spiders. When the eggs hatch, the young wasps eat the animal.

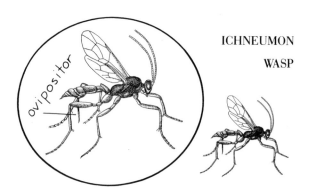

ICHNEUMON
WASP

Bees, wasps, and ants sting when they feel trapped. They sting when you get too near their young. The young may be in a hole in the ground, in a bush, or up in a tree. Sometimes you can see a nest, or hive, made of mud or paper or wax. Sometimes you do not see it in time.

Honeybees, hornets, and most ants live in groups. Most other bees and wasps live alone. All take good care of the young.

A honeybee gets food from flowers

for the queen bee, the young, and her-self. She fans them with her wings to keep them cool on hot days.

A hornet takes home insects and spi-ders. It chews them up to feed the young. A yellow jacket may take home part of your sandwich.

If a *bee* ever stings you, do not rub the spot. Lift the stinger out side-ways with your fingernail. Then squeeze the poison out. With a *wasp* or *hornet*, you need only to squeeze.

☐ Fat cicadas (SIH·KAY·DAS) sometimes bang into windows at night.
They have wide heads and short antennas and sharp beaks. Their eyes seem
to pop from the sides of their heads.
You can see through their big wings.

Mostly, cicadas stay high in the trees
where you cannot see them. You can
hear them buzzing louder and
LOUDER and softer and softer.

The quiet aphid (AY·FID) looks like a
very tiny cicada, but it may not have
wings. You find aphids on the undersides of many soft new leaves and
stems. They are so busy sucking sap
that they do not even run from you.

Along with their hopper cousins,
aphids do much harm. They give
plants germs that make them sick.

Leafhoppers feed in plain sight.
Some have bright stripes and are easy
to see. They are not so easy to catch.
Because they are small and quick, leafhoppers slip through most nets—and
your hands.

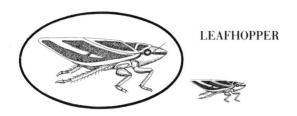

LEAFHOPPER

The spittle insect is easy to catch, but it is not easy to see. It looks like a grayish brown leafhopper. Its young hide in "spit," which they make as they feed. The spit keeps their soft bodies cool and damp on a hot day.

A treehopper looks like a thorn or like tree bark. But if you touch it—zip!—it is gone.

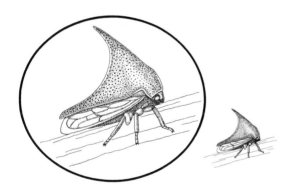

G The front wings overlap. Go to H.

G The front wings do not overlap. Go to I.

Mini-Key G

Mini-Key H

H The front wings cross near the middle. Read below.

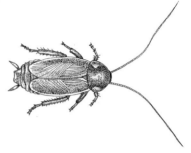

H The front wings cross at the tips. Read below.

☐ Cockroaches slide quickly into cracks. Their flat bodies and strong legs make that easy.

Cockroaches have heavy front wings with many veins. The front wings may be long and overlap in the middle. They may be short and not even touch.

■ A bug is an insect with a broad head, a sharp beak, long antennas, and popping eyes. It has a hard, pointy cover between its flat wings. The front

72

wings are hard and thick at their front ends, but papery thin and clear at the back ends. The back ends cross, looking like one wing tip.

A bug flies mainly with its hind wings. Shorter and wider than the front ones, they may shine like rainbows. When the bug lands, it looks plain again—gray-green or brown.

Most bugs feed on sap. They sink their beaks into a stem or leaf and suck until the plant is black and dry. Some bugs, like the one shown here, stab other insects instead and suck *them* dry. One may stab you by mistake.

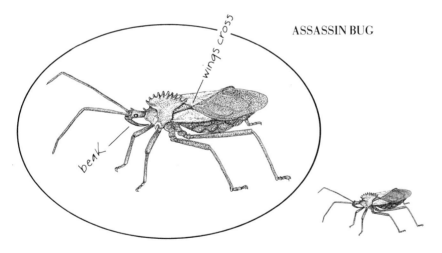

ASSASSIN BUG

The stinkbug and most other bugs make a bad-smelling juice. Berries taste bad even if bugs only walk over them. Toads spit bugs right out.

Mini-Key I   **☐** The front wings cover the abdomen. Go to J.

■ The front wings do not cover the abdomen. Read below.

■ Insects from four groups look much alike. All are brown or black. All run about at night. All chew their food.

You can tell one from another by looking at the antennas and the cerci.

ANTENNAS	CERCI	INSECT
short	none	 see page 77
long	long	 see page 64
very long	short	 see page 72
long	long, like hooks or straight	 read below

Earwigs have short, thick front wings. They use their cerci to fold their bigger hind wings under them. Young earwigs

74

and some females do not have wings. Their cerci are straight. The males may fight with their cerci.

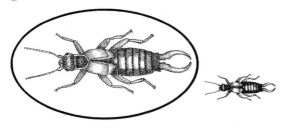

J The front wings slope up like the sides of a pup tent. Look again:

- The insect is small, with short, stiff antennas and a beak. Turn to page 71.

- The insect is big, with long, hairlike antennas and a flat face. Turn to page 64.

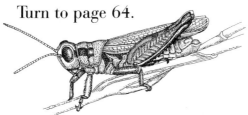

J The front wings make a tight, waxy shell over the hind wings. Turn to the next page.

Mini-Key J

A ladybug is round. A firefly is long. Each is a beetle.

Beetles come in many sizes and shapes. They come in many colors, with pretty lines and spots. Most beetles, though, are plain black or brown. Their thick front wings may be shiny and smooth, or rough and bumpy. Most are long and cover the abdomen tightly.

When a beetle takes off, it holds its front wings to the sides. It flies with its thin hind wings only. The hind wings are bigger than the front ones. They fold up neatly under the front wings when the beetle is not flying.

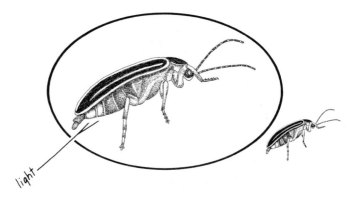

light

Beetles are strong fliers. Yet they fly very little, even when you try to catch them. Instead, they try to make *you* go away.

Some beetles give off yellow drops, which may taste bad to birds. Some buzz loudly, like bees. Others wave their antennas or snap their jaws. Some beetles pull their heads in and play dead. Others let themselves fall to the ground and then start running.

The rove beetle runs very fast. Its wings are short. It holds the tip of its abdomen up. It looks ready to sting, but, really, it has no stinger.

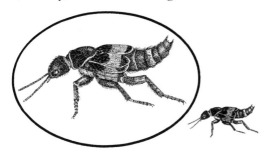

Beetles can bite and chew, but they cannot sting. A large beetle can bite you if you hold it too tight.

The beetles you have seen so far are "good" insects. They eat animals that harm plants that people want. For example, fireflies eat slugs and snails. So do their young, which are called glowworms. Ladybugs eat aphids.

Ground beetles eat maggots and caterpillars. Scarab beetles, like the one below, clean up dead plants and animals. But they also eat leaves and fruit.

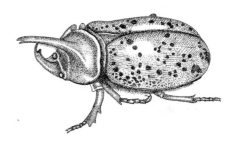

Many other beetles though are pests. Their young are pests, too. Their strong jaws can make long holes in trees and shrubs. Some kinds eat vegetables or fruits. Others eat grain or other crops that people need. Even a tiny beetle can chew so many holes that soon there are more holes than leaf.

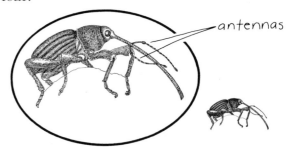

antennas

Some of the worst pests are called weevils, or snout beetles. Their heads are mostly snout. The long snout can twist into tight places. The jaws are at its tip. They can drill even into hard wood.

With weevils, both grown-ups and young hurt the plants they eat. (To see what the young look like, turn to page 80.)

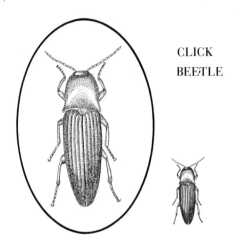

CLICK
BEETLE

The young of other beetles are bigger pests than the grown-ups. This is true, for example, of young click beetles, or wireworms. The click beetle gets its name from the noise it makes. If you lay one on its back, it bends its body and, *snap*, it flips over. Sometimes it misses and has to try again.

MAGGOTS AND GRUBS

These soft, pale animals are called **maggots** or **grubs**. Sometimes they are called borers or worms or miners. While they are young, they have no legs.

This one has no head. It is a young fly.

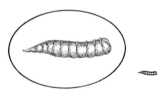

This one has a head, but no eyes and no feelers. It is a young weevil.

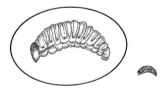

They do not need legs or eyes or feelers—they only need mouths. A maggot or grub just wiggles to get its next bite.

Its parents have made sure there is food. Some may bring food to their young or carry their young to food. Many parents lay their eggs on food the young will eat.

Food may be the soft inside of a leaf or stem or root. It may be a cotton or rice flower. It may be a peach or a plum. Food may be a dead animal or even a *live* one.

Before the food is gone, maggots and grubs change into pupas (P·YOU·PAS). (See page 84.) The pupa looks like the parent animal wrapped in brown plastic.

At last a grown-up animal comes out. Now you know what the soft, pale animal was. (To see the weevil, turn to page 79. To see the fly, turn to page 58.)

CATERPILLARS

One day eggs are shining on a leaf like specks of gold or green. Soon some tiny, wormlike animals hatch. They have many legs and big mouths. They are **caterpillars**, ready to munch on plants.

The inchworm caterpillar has ten legs. Its six front legs have five joints. (Here's a hint: The legs of its parents do, too.) Each leg ends in a tiny claw.

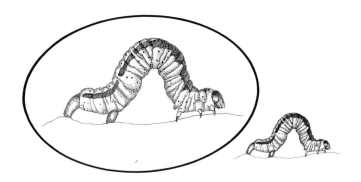

The four back legs have no joints. Soft and wrinkly, each one stands on tiny hooks. The soft legs help the hind end catch up to the head end.

82

The woolly bear caterpillar has six jointed legs and ten soft legs. Few caterpillars have more than ten legs without joints.

You can feel how the legs with joints differ from the legs without joints just by letting a caterpillar walk on your hand. *But please do not try this with a hairy one.* The hairs give some people rashes. A mole may eat the caterpillar in spite of the hairs.

A caterpillar cannot run off or bite. Still, holding one is not easy. It may try some tricks to get away. It may hump up its back. It may jerk the front part of its body into the air or sway like a snake. Does it scare a hungry bird?

Caterpillars do little but chew. In a few days their tough skin cannot stretch anymore. Then the caterpillar sheds its old skin and crawls out in a soft, new one. Before the new skin gets stiff, the caterpillar puffs itself with air.

The wrinkly new skin becomes smooth. There is now room inside for growing.

The young caterpillar goes on eating.

HORNWORM CATERPILLAR

head

Soon its new skin is too tight, and it sheds again. Each time it may change color. It may grow horns or knobs or hairs. It may get new stripes or spots. Such things help it match the plants it is eating—or scare away an enemy.

The caterpillar grows as big as its parents, and then even a bit bigger. When it is fully grown, it grows a hard skin around itself and becomes a pupa (P·YOU·PA). Some caterpillars may first wrap themselves in a cocoon (KUH·COON) of silk and hairs. They may just bury themselves in the ground. Other caterpillars make cases that match the plants around them.

The word *pupa* means "doll," and, like a doll, the pupa does not move. Its back part, or abdomen (AB·DOH·MEN), may twitch if you touch it. Inside the hard cover, the animal is changing. It loses its wormlike shape and its soft legs. It gets new mouth parts and it gets wings.

The changes may happen quickly or

slowly. Each animal changes at its own rate and in its own time.

At last the animal pushes out of its case and slowly spreads its wings. The wings are soft and flabby. The animal gives them little shakes as blood moves through them. Now it is ready to fly.

Here is the inchworm, grown up.

Here is a grown-up woolly bear.

OTHER BOOKS ABOUT
CREEPY CRAWLIES

Borror and White, *A Field Guide to the Insects of America North of Mexico.* The Peterson Field Guide Series. Boston, 1970.

Klots, *A Field Guide to the Butterflies of North America, East of the Great Plains.* The Peterson Field Guide Series. Boston, 1951.

Levi and Levi, *A Guide to Spiders and Their Kin.* A Golden Nature Guide. New York, 1968.

Pringle, *The Hidden World: Life Under a Rock.* New York, 1977.

Pringle, *Twist, Wiggle, and Squirm: A Book About Earthworms.* New York, 1974.

Simon, *Snails of Land and Sea.* New York, 1976.

Stokes, *A Guide To Observing Insect Lives.* Boston, 1983.

Swain, *The Insect Guide.* Doubleday Nature Guide Series. New York, 1948. *Hard to find.*

White, *A Field Guide to the Beetles of North America.* The Peterson Field Guide Series. Boston, 1983.

INDEX

The index lists all the animals in this book by name, with the pages where you can read about them. It also tells you, in heavier type, the page where you can see what the animal looks like and the real-life size of the animal, both in millimeters (mm) and in inches (in). When one measurement is given (25mm/1in), it is for the animal shown. When a range is given (3 – 10mm/⅛ – ⅜in), it is for the animal's group.